Ramon Cruz Borges
Francisca Robevania Medeiros Borges
Rebeca de Araújo Torres

Evaluation of an irrigation system in areas cultivated with pastures

Ramon Cruz Borges
Francisca Robevania Medeiros Borges
Rebeca de Araújo Torres

Evaluation of an irrigation system in areas cultivated with pastures

Mesh sprinkler irrigation in areas cultivated with brachiaria and tifton grass

ScienciaScripts

Imprint

Cover image: www.ingimage.com

This book is a translation from the original published under ISBN 978-3-330-75990-9.

Publisher:
Sciencia Scripts
is a trademark of
Dodo Books Indian Ocean Ltd. and OmniScriptum S.R.L publishing group

120 High Road, East Finchley, London, N2 9ED, United Kingdom
Str. Armeneasca 28/1, office 1, Chisinau MD-2012, Republic of Moldova, Europe
Printed at: see last page
ISBN: 978-620-8-31413-2

CONTENTS

Evaluation of irrigation systems in areas cultivated with pastures

SUMMARY

Evaluating the efficiency and uniformity of irrigation systems at field level helps to mitigate the losses that occur during and after the application of water by the system. The aim of this study was to evaluate the performance of a mesh sprinkler irrigation system installed in a pasture area with Brachiaria and Tifton 85 grass in the municipality of Fortaleza - CE. The system's water application uniformity was estimated using the Christiansen Uniformity Coefficient (CUC), the Distribution Uniformity Coefficient (CUD) and the Statistical Coefficient (CUE). On the basis of the values found, it can be concluded that the irrigation system evaluated is impaired, since all the performance parameters evaluated showed low values, classified as poor when compared to those recommended in the literature.

Keywords: Coefficient of Uniformity, *Brachiaria decumbens*, *Cynodon spp.*

1 INTRODUCTION

In recent years, due to climate instability with a direct impact on rainfall distribution, the evolution of Brazilian livestock farming is basically due to the participation of irrigation in agribusiness and among the irrigation methods most used in pastures, pressurized systems stand out, especially sprinkler irrigation systems, the most widely used of which are center pivot and mesh sprinkler (DRUMOND; AGUIAR, 2005).

According to Drumond and Fernandes (2004), in a mesh sprinkler irrigation system, the lateral lines are interconnected in a mesh and buried, as are the main and branch lines. Its main characteristics are: low energy consumption, the possibility of dividing the area into several sub-areas, ease of operation and maintenance, the possibility of fertigation and low installation and maintenance costs.

In any irrigation project, it is necessary to check whether the conditions initially planned are being met in the field. The evaluation must take into account pressure, flow, applied water levels and application uniformity, among other parameters (MANTOVANI et al., 2009). However, evaluating the performance of irrigation systems is still infrequent on most rural properties and when an evaluation is carried out, it is usually done without any advice and, in general, fails to return the expected increases in plant material production to the producers.

The rational management of any irrigation project must take into account social and ecological aspects and seek to maximize productivity, minimize costs, increase efficiency in the use of water and energy, while maintaining soil moisture conditions favourable to good plant development (PAZ et al., 2000).

Irrigated agriculture has many benefits that can only be fully achieved when the irrigation system is used with management criteria that result in quality water applications, at the right time and in quantities compatible with the consumption needs of the irrigated crops, always taking into account the soil-plant-water-climate relationship (DRUMOND, 2003).

In view of the above, the aim of this study was to evaluate the irrigation quality of a mesh sprinkler system, operating under field conditions, by quantifying performance parameters in an area cultivated with *brachiaria* grass (*Brachiaria decumbens*) and tipton 85 grass (*Cynodon ssp.*) located at the NEEF/DZ/CCA/UFC in Fortaleza - CE.

2. LITERATURE REVIEW

2.1 Economic aspects of pasture irrigation in Brazil

Irrigation can be defined as the artificial application of water to the soil to provide adequate moisture for the normal development of cultivated plants, in order to make up for the lack or poor distribution of rainfall.

Irrigated agriculture has been an important strategy for improving global food production, providing sustainable development in the countryside and generating jobs and income in a stable way. Currently, more than half of the world's population depends on food produced in irrigated areas (MANTOVANI et al., 2009).

Between 1996 and 2006, the Brazilian cattle herd increased from 158.3 to 205.9 million animals, while the total area of pasture decreased from 177.7 million to 158.6 million hectares (IBGE, 2010). These figures reveal the efforts made by technicians, producers and researchers over this period to increase pasture productivity, which has resulted in a considerable improvement in pasture utilization efficiency (EUCLIDES et al., 2010).

Even with this development, the average productivity of Brazilian pastures is still incipient, with average stocking rates of less than 1 animal unit (AU) per hectare, indicating a need to increase production (ALENCAR et al., 2009). When thinking about increasing pasture production, research has sought the rational use of technologies related to the management of the soil, the environment, the plant and the animal. According to the authors, these technologies include the rotational grazing method, the use of fertilizer levels compatible with high forage production and the use of irrigation.

In this situation, the use of irrigation in Brazilian pastures is progressive, particularly in the warmer regions, which show greater responses to its use (DRUMOND; AGUIAR, 2005). However, the use of this technology must be combined with the appropriate use of fertilizers, so that the actual increase in production is in line with what is expected.

According to Garcia et al. (2007), the semi-arid region of northeastern Brazil is usually affected by an irregular period of rainfall, due to its insufficiency and poor distribution over the course of the year, in such a way that this constitutes one of the most restrictive factors for agricultural productivity, considerably compromising crop yields. These facts highlight the need to resort to irrigation in order to alleviate and correct the problems caused by inefficient water supply in agriculture.

Bilibio et al. (2010) report that in irrigated agriculture, special attention must be paid to irrigation management, since it is essential to rationally manage the irrigated crop under field conditions, accurately demonstrating the crop's water needs without deficit or excess, as well as the most appropriate rate and time to irrigate, in order to maximize water use efficiency.

Irrigation should not be considered in isolation, but as part of a set of techniques used to guarantee the economic production of a given crop with appropriate management of natural resources. Therefore, aspects such as planting systems, crop rotation possibilities, soil protection, soil fertility, integrated pest and disease management, mechanization, etc. should be taken into account, in pursuit of integrated production and better market insertion (MANTOVANI et al., 2009).

2.2 Irrigation systems

The use of irrigation systems has been an important strategy for optimizing food production by increasing the productivity and quality of a crop, favoring off-season production and generating employment and income in the countryside (LUNA et al., 2013).

According to Andrade (2001), some factors must be taken into account when deciding whether to irrigate crops, such as the amount of water available, the cost/benefit of using this technique and how/how much rainfall is distributed in the region. The most important factor in deciding whether to irrigate a crop is the distribution of rainfall, because during the production of a given vegetable, there are periods of greater demand for water and others of lesser demand, which requires great attention from the producer.

Seasonal water demand, for example, is the amount of water that a crop uses during its cultivation. Within the growing cycle, there is a critical period when water demand is greater than normal (CARVALHO et al., 2013). Thus, irrigation appears as an alternative to meet the demands of the producer. It is a technique used with the aim of balancing the water deficit during periods of seasonal water demand, as well as critical periods of water demand.

The main irrigation methods used are: surface irrigation, sprinkler irrigation, localized irrigation and sub-irrigation (ANDRADE; BRITO, 2006). Each method has different costs, advantages and limitations and in the vast majority of cases is suitable for different crops.

2.2.1Surface irrigation

In this method, water is distributed by gravity through the soil surface. It requires a large volume of water to manage. It is a method that requires uniform areas, without elevations or depressions, to avoid water shortages or accumulations (BERNARDO et al., 2009).

2.2.2Sprinkler irrigation

Water distribution in this method is by means of water jets which, when released into the air, fall on the crop in the form of rain. It is a method that adapts to almost all crops, although it interferes with cultivation by washing the aerial part of the plants. It adapts to various soil and climate conditions. It can be fully automated and portable. It is influenced by wind conditions and relative humidity, which affect distribution uniformity. The use of saline water is not recommended, as it reduces the useful life of the equipment and causes possible damage to plant leaves (BERNARDO et al., 2009).

According to Andrade and Brito (2006), the sprinkler irrigation method in Brazil is divided into five systems: a) conventional sprinkler; b) self-propelled; c) center pivot; d) linear displacement; e) LEPA.

i) *Conventional Sprinklers - These* can be fixed, semi-fixed or portable systems. In fixed systems, the pipes are permanently buried, so less labor is required. In semi-fixed and portable systems, labor is required to change the lateral lines. These systems are recommended for small properties, using family labor to reduce installation costs.

ii) *Self-propelled* - In this system, a cannon or mini cannon is mounted on a trolley that moves longitudinally along the area to be irrigated. The pressure of the water drives the trolley. This is the system that consumes the most energy and is heavily influenced by the wind, which interferes with the distribution of water in the irrigated area.

iii) *Central Pivot* - Consists of a lateral line that rotates around a circle (pivot). These systems allow a high degree of automation and are used to irrigate large areas.

iv) *Linear* displacement - The lateral line has a similar displacement to the center pivot, but moves in a transverse position and in the longitudinal direction of the area. This system is recommended for flat rectangular areas with no obstructions.

v) *LEPA "low energy precision application"* - These are center pivot or linear displacement systems, containing equipment with more efficient water application mechanisms, in which the water is applied directly to the soil surface, which reduces water losses by wind and prevents the plant from getting wet.

2.2.3. Targeted Irrigation

This method allows water to be applied to the soil, directly to the root region of the plants, at low intensities but with high frequencies so that the moisture in the soil is close to "field capacity". The emitters used can be drippers (porous tubes) or surface emitters (micro-sprinklers). The wetted area corresponds to 20 to 80% of the total area, favoring water savings. The initial cost of implementing this system can be relatively high, as it allows full automation. It requires less manpower to operate. According to Andrade and Brito (2006), the most commonly used localized irrigation systems are: a) Drip; b) Micro-sprinkler; c) Surface Drip.

In the drip system, the water is applied pointwise to the surface of the soil and does not wet the foliage or the stalks of the plants. This system requires water filtration to prevent the emitters from clogging up. In the case of the subsurface drip system, the porous tubes are buried, allowing the water to be applied subsurface. This system has a longer lifespan and allows the lateral lines to be removed from the surface of the soil to facilitate traffic and cultivation.

2.2.4Subirrigation

In this system, the water table is kept at a certain depth, capable of allowing a flow of water that is beneficial to the root zone of a given crop (ANDRADE; BRITO, 2006). It is usually associated with a surface drainage system.

2.3 Irrigation systems for pastures

In recent years, Brazilian livestock farming has been promoting the use of irrigation in agribusiness, which is an important strategy for increasing production, productivity and profitability on rural properties (BERTOSSI et al., 2013). Among the irrigation methods most commonly used in pastures are pressurized systems, such as sprinkler irrigation (DRUMOND; FERNANDES, 2004).

An efficient irrigation system should sprinkle water evenly over the soil, to a certain depth, favoring the humidity necessary for the normal development of plant species (DRUMOND, 2003).

The types of sprinkler irrigation systems most commonly used in pastures are:

- Conventional sprinkling (semi-portable, portable and fixed);
- Mesh sprinkling (underground, fixed or surface system);

- Central pivot (large areas).

Sprinkler irrigation systems use emitters called sprinklers, which distribute the water to the plants. Sprinklers can have one or more nozzles, mechanisms through which the water that is sprayed onto the plants exits. The pressurized water passes through the nozzles, is sprayed and falls as artificial rain. The working pressure is generally obtained by means of motor pumps (electric or diesel), but it can also result from the unevenness of the terrain (BERNARDO et al., 2009).

Bernardo et al. (2009) report the following advantages of sprinkler irrigation systems: There is flexibility in the rate of water application and it is easy to adapt to the soil's water infiltration capacity and the stage of crop development; There is no need to systematize the land and it reduces the cost of installation and operation in different topographical conditions; There is better use of the land, with no need to use canals and furrows for water conduction; It favours saving labour for water conduction, especially in fixed and mechanized systems; There is less water loss (evaporation and infiltration) due to the transport of water in pipes, compared to surface irrigation; It enables the automation and application of chemical products and fertilizers through irrigation water (fertigation / chemigation) and provides good uniformity of application and high efficiency of water distribution in the irrigated area, if the system is well dimensioned and well managed.

Mendonça et al. (2007) state that conventional sprinkler irrigation systems are more widely used in family farming because of the small size of the areas to be irrigated, the main systems being semi-portable and fixed. They also highlight the mesh sprinkler system as being the most practical and least expensive to install. These generally have fixed pipes and movable sprinklers, with the pipes being buried, thus increasing their useful life.

According to Drumond (2008), the mesh sprinkler system has been gradually expanding in Brazil due to its practicality, ease of modulation and adaptability to the investment capacity of each owner.

Irrigation of pastures using center pivots gained market share in the 90s. Irrigated pastures originated in areas of agricultural grain production, where there was low productivity caused mainly by degraded soils contaminated by pathogens (CARDOSO, 2001). According to the same author, costs are highly relevant in center pivot irrigated systems. In addition, not all rural properties have the topographical conditions, electricity, water and soil to adopt this technology, making it unfeasible to implement it in regions where the water deficit caused by the scarcity of rainfall is quite severe.

2.4 Evaluation of the irrigation system

The concept of irrigation efficiency encompasses two basic aspects: the uniformity of application and the losses that can occur during the system's operating period. In order for efficiency to reach acceptable levels, it is necessary for losses to be as low as possible during the system's operation.

According to Barreto Filho et al. (2000), the uniformity of water application is a parameter that details the irrigation system as a function of the difference in the volume sprayed on the plant along the lateral lines, with a direct result on crop performance, which is why it is considered one of the most decisive factors in the design and execution of irrigation systems.

Several coefficients are used to demonstrate the variability of the distribution of water applied by a sprinkler irrigation system on the soil surface. The first of these was suggested by Christiansen (1942) and uses the mean absolute deviation as a measure of dispersion (CUC). Wilcox and Swailes (1947) suggested a coefficient of uniformity using the standard deviation as a measure of dispersion, for which values above 75% are allowed (CUE); now Criddle et al. (1956) have inserted another measure of uniformity, considering the ratio between the average of the lowest quartile and the average blade collected (CUD).

Water distribution uniformity (WDU) is used as an indicator of irrigation distribution problems. A low CUD value indicates abundant water loss through deep percolation, when the minimum water applied corresponds to the actual irrigation water required, where CUD values are generally lower than CUC.

According to Mantovani (2001), the values of the uniformity coefficients (CUC, CUD and CUE) were organized into a classification scale as shown in Table 1.

Table 1 - Classification of the performance values of conventional sprinkler irrigation systems according to the Christiansen Uniformity Coefficient (CUC), the Statistical Uniformity Coefficient (Us) and the Distribution Uniformity Coefficient (CUD).

Classification	CUC	CUD
		(%)
Excellent	>90	>84

Good	80-90	68-84
Fair	70-80	52-68
Bad	60-70	36-52
Unacceptable	<60	<36

Source: Mantovani (2001).

The acceptable CUC value adopted as a reference by convention for sprinkler irrigation systems is 0.85 or 85%. According to Gomes (1994) CUC values < 80% can be accepted if: (i) rainfall is significant during the growing season, (ii) the cost of the system is small enough to compensate for the reduction in net income due to the decrease in production and (iii) the Iateral lines are operated in alternating positions during successive irrigations.

The American Society of Agricultural Engineers (ASAE), through its Committee on Sprinkler Irrigation, presented a table of recommendations for a correct comparison of uniformity referring to the ASAE CUE (1966) which is shown in Table 2.

Table 2 - Classification of the Statistical Uniformity Coefficient (SUC).

Degree of Acceptability	**CUE**
Excellent	100-95
Good	90-85
Normal	80-75
Bad	70-65
Unacceptable	<60

Source: ASAE (1996).

3. MATERIALS AND METHODS

3.1 Experiment site, soil type and climatic conditions

The experiment was conducted in November 2015 in four paddocks, identified by the numbers three (3) and six (6), which were cultivated with Tifton 85 grass, and paddocks eleven (11) and fourteen (14), cultivated with brachiaria grass, both belonging to the experimental area located at the NEEF/DZ/CCA/UFC, Campus do Pici, in the municipality of Fortaleza - CE.

The municipality of Fortaleza is located in the coastal zone at an altitude of 15.49 m, latitude 30°43'02" south and longitude 38°32'35" west. The soil in the study area is of the Argissolo Vermelho Amarelo type (EMBRAPA, 2006). According to the Koppen classification, the area is located in a region with an Aw' climate, which is characterized by a rainy climate, with precipitation from fall to summer and an average temperature of over 18°C throughout the year.

3.2 Procedure in the field

The first step was to choose a location in the field for the test along the lateral sprinkler line. The evaluation of the system already installed was carried out in the middle 1/3 of the lateral line, according to the methodology adopted by Bernardo et al. (2009), who report that level lines have a better representation of the average operating pressure of the sprinklers. These sprinklers are generally located at a distance of 40% of the total length from the start of the line.

The system evaluated was a conventional mesh sprinkler, with a NaanDanJain® model

435 ½^{,5} M sprinkler, with a complete and sectorial rotation, which according to the manufacturer's specifications has a wetted diameter of 23 to 29 m with a flow rate of 0.38 to 1.18 m h$^{3,-1}$ and a working pressure of 1.5 to 4 bar (Figure 1). The equipment used to evaluate the irrigation system consisted of Fabrimar collectors, a measuring cylinder in mL, a stopwatch, a Pitot tube, a container with a capacity of 6.0 L, a hose, a measuring tape and a portable thermo-anemometer.

Figure 1 - NDJ sprinkler model 435 ½^{,5} M, Fortaleza - CE, 2016.

Photo: Borges, R. C.

Thus, the distances S1 (sprinkler spacing) and S2 (lateral spacing) of the four paddocks evaluated were measured, dividing these areas into m^2 . To demarcate the area, rain collectors spaced out in 2 x 2 m squares were used with the aid of a measuring tape. In the center of each sub-area, a metal rod was fixed to the ground for each collector, positioned at an average height of 60 cm from the ground surface, which served as a support for the collectors. These were placed between six sprinklers, making a total of 21 units in Area 3, 21 units in Area 6, 24 units in Area 11 and 36 units in Area 14, as shown in the sketch in Figure 2.

Figure 2 - Sketch of the areas evaluated in the experiment. Fortaleza - CE, 2016.

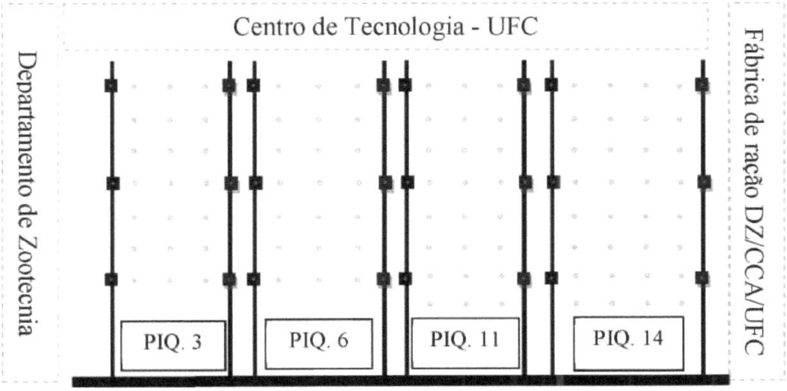

Fonte: Borges, R. C.

LEGENDA:
Coletor pluviométrico
Aspersor

The collectors were placed in an inverted position to prevent water from collecting before the evaluation began. The centralization of the collectors was checked and, once everything was correct, the motor-pump unit was started. We waited a few minutes until the system's pipes were completely pressurized.

The sprinklers' flow rates were measured before the tests were carried out using a hose attached to the nozzle of each sprinkler, a reservoir with a capacity of 6.0 L and a precision chronometer. The flow rate determined was the arithmetic mean of three repetitions for each sprinkler in each picket.

With the help of a Pitot tube coupled to a Bourdon pressure gauge, the working pressure of the 6 sprinklers in the areas under evaluation was checked (Figure 3). The time taken for the sprinkler to travel through each sector and complete its full rotation was then checked with a stopwatch. The time taken represented the arithmetic mean of three repetitions for each sprinkler evaluated.

Figura 3 - Verification of sprinkler working pressure using a Pitot tube coupled to a Bourdon manometer. Fortaleza - CE, 2016.

Photo: Borges, R. C.

The motor pump was switched off and the sprinklers stopped working. The collectors were turned upwards and plugged into the stem so that the test could begin. When the evaluation began, each collector captured all the sprinkled water (Figure 4). The duration of each collection was approximately half an hour, corresponding to the irrigation time used by the irrigators in the area.

Figura 4 - Layout of collectors for water collection. Fortaleza - CE, 2016.

Photo: Borges, R. C.

While the irrigation system was running, the temperature and wind speed were measured at precise 10-minute intervals using a portable thermo-anemometer (Figure 5).

Figura 5 - Measuring air temperature and wind speed using a thermo-anemometer. Fortaleza - CE, 2016.

Photo: Borges, R. C.

The evaporation determined during the test was measured using a known volume of water placed in four different collectors (Figure 6), and at the end of the test the difference between the initial volume and the final volume of the four collectors was taken.

Figura 6 - Arrangement of collectors in the field to determine evaporation.

Fortaleza - CE, 2016.

Photo: Borges, R. C.

At the end of the test, the volumes of water collected were measured using a Fabrimar measuring cylinder graduated in mm.

3.3 Parameter evaluations

The following parameters were used to evaluate the performance and efficiency of the conventional sprinkler irrigation system: Christiansen Uniformity Coefficient (CUC), Distribution Uniformity Coefficient (CUD), Statistical Uniformity Coefficient (CUE), Irrigation Efficiency (EI), Water Distribution Losses (PDA), Potential Application Efficiency (EAP) and evaporation and wind drag losses (Pv).

Christiansen (1942) applied the concept of measuring the uniformity of water distribution by a sprinkler system, calling this method the "coefficient of uniformity" (CUC), expressed by equation 1:

$$CUC = 100\left[1 - \frac{\sum_{i=1}^{n}\left|X_i - \overline{X}\right|}{n \cdot \overline{X}}\right] \quad (1)$$

in which,

CUC - Christiansen Uniformity Coefficient, in percentage; n - number of collectors;

X_i- precipitation height collected at the i-th collector;

$\bar{X}$ - average blade applied.

The Coefficient of Uniformity of Distribution (CUD) was calculated using equation 2:

$$CUD = 100 \cdot \frac{\overline{x}}{\overline{X}} \quad (2)$$

in which,

$\bar{x}$ - average rainfall of 25% of the rain gauges with the lowest rainfall;

$\bar{X}$ - average rainfall, considering all the rain gauges.

The coefficient of statistical uniformity (CUE) is another parameter that can be used to determine the uniformity of irrigation systems and was calculated using equation 3:

$$CUE = 100\left(1 - \frac{S}{\overline{X}}\right) \quad (3)$$

in which,

S - standard deviation of precipitation data;

$\bar{X}$ - average rainfall, mm.

The CUE value considered satisfactory is between 0.75 and 0.80 (75 and 80%).

The standard deviation of the rainfall data was determined using equation 4:

$$S = \sqrt{\frac{\sum_{i=1}^{N} \left| X_i - \overline{X} \right|^2}{N}} \tag{4}$$

in which,

X_i - blade observed in collector i, mm;

$\bar{X}$ - average rainfall, mm;

N - number of collectors.

The Irrigation Efficiency (IE) parameter is used to determine how efficiently a sprinkler system can use the water supply. It is defined by the ratio between the 25% smallest blades or volumes applied and the average blade applied (L), obtained from the following equation (5):

$$EI = \frac{\bar{x}}{L} \cdot 100 \tag{5}$$

in which,

$\bar{x}$ - average rainfall of 25% of the rain gauges with the lowest rainfall;

L - average spray applied, L = PPT sprinkler x irrigation time.

The Water Distribution Losses (WDP) of the system evaluated were obtained using equation 6:

$$PDA = 100 - EI \tag{6}$$

where, PDA - Water loss per application (%);

EI - Irrigation efficiency (%).

According to Mantovani et al. (2009), losses due to evaporation and wind drag increase with air temperature, wind speed and, consequently, with the sprinkler's operating pressure. Potential application efficiency can be determined by equation 7:

$$Eap = \frac{Lm \cdot S1 \cdot S2}{q \cdot ti} \cdot 100 \qquad (7)$$

in which:

Eap = potential application efficiency, %; q = average sprinkler flow, $l\ s^{-1}$;

ti = irrigation time, s;

Lm = Average blade collected, mm;

S1 = sprinkler spacing, m;

S2 = spacing between sides, m.

The loss due to evaporation and wind drag (Pv) can be determined by the following equation 8:

$$Pv = 100 - Eap \qquad (8)$$

4. RESULTS AND DISCUSSION

4.1 Sprinkler technical parameters

Table 3 shows the technical parameters of the sprinklers in the four paddocks evaluated.

Table 3 - Technical parameters for the sprinklers evaluated: Working pressure (Ps), Flow rate, Application intensity (Ia) and Rotation per minute (RPM) in areas with pastures already installed. Fortaleza - CE, 2016.

	Picket -3 ---------------------------				--- Picket 6 -----------------			
Sprinkler	Ps (mca)	Flow (L h)$^{-1}$	Ia (mm h)$^{-1}$	RPM	Ps (mca)	Flow (L h)$^{-1}$	Ia (mm h)$^{-1}$	RPM
1	15,5	563,6	6,26	9,98	16,0	589,2	7,20	8,57
2	16,0	563,8	6,26	10,96	16,0	594,6	7,27	10,00
3	15,0	531,2	5,90	9,13	16,0	583,8	7,14	8,57
4	18,0	617,1	6,86	11,79	19,0	648,1	7,92	8,57
5	17,8	623,5	6,93	6,73	19,0	635,3	7,77	10,00
6	17,5	589,2	6,55	12,24	19,0	641,7	7,84	7,83
Average	16,6	581,4	6,46	10,14	17,5	615,5	7,52	8,92
		----Picket 11 ----				--- Picket 14 ---		
Sprinkler	Ps (mca)	Flow (L h)$^{-1}$	Ia (mm h)$^{-1}$	RPM	Ps (mca)	Flow (L h)$^{-1}$	Ia (mm h)$^{-1}$	RPM
1	15,0	544,6	5,98	10,19	21,0	668,2	5,14	8,57
2	15,0	558,7	6,13	10,42	21,0	661,4	5,09	9,00
3	14,5	553,8	6,08	9,17	20,0	641,7	4,94	7,83

4	17,0	573,5	6,30	8,66	22,5	696,8	5,36	8,18
5	16,0	600,0	6,59	9,47	22,0	682,3	5,25	7,50
6	16,0	573,5	6,30	8,16	21,5	648,1	4,99	5,63
Average	15,6	567,4	6,23	9,35	21,3	666,4	5,13	7,78

Source: Borges, R. C., 2016.

When analyzing these parameters, it can be seen that *there are* variations between the highest and lowest sprinkler flow rates, with a variation of 17.4% in paddock 3 and 11% in paddock 6. In paddock 11, the variation was 10.2% and finally paddock 14 showed a variation of 8.6%.

With regard to sprinkler flow rates, it can be said that all of them are in accordance with the manufacturer's catalog, ranging from 531.2 to 696.8 L h^{-1} . The intensity of application varies between the highest and lowest values: 17.5% in paddock 3, 10.9% in paddock 6, 10.2% in paddock 11 and 8.5% in paddock 14. Even though it varies, the sprinkler operating pressure is within the range recommended by the manufacturer.

According to Alencar et al. (2009) when an irrigation project has sprinklers operating at different working pressures, different flow rates will be applied and consequently lower water application uniformities will be observed. As in this study the pressures of the sprinklers evaluated were different, there was a high variation in the flow rate applied.

There was also a big difference between the rotations of the sprinklers in the four paddocks, which operated in the range of 5.63 to 12.24 RPM (revolutions per minute), This is proof of the high divergence between the rainfall collected during the test and also explains the low performance values found in the irrigation of the areas in question, since measuring the rotation times of the sprinklers in the quadrants indicated that the water was discharged with uneven distribution in all directions.

Bernardo et al. (2009) consider that the rotational speed of the sprinkler should vary from 0.5 to 2.0 RPM. All the sprinklers showed values higher than those admissible for sprinkler irrigation to be conducted properly. According to Christiansen (1942), high sprinkler rotational speed leads to a reduction in the area covered by the emitter, resulting in a corresponding increase in the actual application rate of up to 40% when compared to the same sprinkler rotating at its recommended speed.

The spacing between sprinklers is considered to have a major influence on the uniformity of water distribution in the area to be irrigated. Table 4 shows the spacing between sprinklers (S1) and between lateral lines (S2). It can be seen that there is a certain irregularity between the spacings in both S1 and S2, which certainly adds to the other factors discussed in contributing to the low uniformity of rainfall over the area under evaluation.

Table 4 - Spacing S1 (spacing between sprinklers) and S2 (spacing between lateral lines) of the sprinklers installed in the paddocks evaluated. Fortaleza- CE, 2016.

PIQUETTE 3 -------------------------				**PIQUET 6 ----------------------------**			
Sprinkler no.	S1 (m)	Sprinkler no.	S2 (m)	Sprinkler no.	S1 (m)	Sprinkler no.	S2 (m)
1-2	6,00	1-4	6,10	1-2	6,27	1-4	6,20
2-3	6,00	2-5	6,15	2-3	6,25	2-5	6,30
4-5	6,10	3-6	6,05	4-5	6,25	3-6	6,15
5-6	6,20	-	-	5-6	6,25	-	-

-- PIQUET 11 --------				--- PIQUETTE 14 -----			
Sprinkler no.	S1 (m)	Sprinkler no.	S2 (m)	Sprinkler no.	S1 (m)	Sprinkler no.	S2 (m)
1-2	6,72	1-4	6,85	1-2	7,60	1-4	7,60
2-3	6,76	2-5	6,80	2-3	7,60	2-5	7,60
4-5	7,00	3-6	6,90	4-5	7,40	3-6	7,60
5-6	6,15	-	-	5-6	7,70	-	-

Source: Borges, R. C., 2016.

Faria et al. (2009) confirm that uniformity in conventional sprinkler systems results from factors such as the type of sprinkler and its operating conditions, the layout and spacing between sprinklers in the field, wind speed and direction during the water application period.

The water distribution characteristics of sprinklers come from the sprinkler model considered and its operating conditions such as: installation height, nozzle diameters, working pressure, jet exit angle and rotation speed, as well as the climatic conditions during operation (KING et al., 2011).

Studies carried out with the Plona RL-250 gun have shown that the wind speed between 0 and 5 $m\ s^{-1}$ has a linear relationship with the reduction in the wetting radius in the perpendicular and opposite directions to the wind, and an increase in relation to the direction of the wind (OLIVEIRA et al., 2009). The effect of the wind can be minimized by decreasing the spacing along the side rows and between side rows (BERNARDO et al., 2009).

Knowledge of the water distribution characteristics of sprinklers is essential for designers to establish the best operating conditions for sprinkler irrigation systems in order to optimize water application uniformity (PRADO et al., 2007).

4.2 Irrigation system performance indicators

The values of the coefficients calculated from the uniformity test carried out for the irrigation system can be seen in Table 5. With regard to water distribution uniformity, it was found that the Christiansen Uniformity Coefficient (CUC) and the Distribution Uniformity Coefficient (CUD) showed values below the good performance range, according to the classification mentioned in Table 1, in paddocks 3, 6 and 14, with the exception of paddock 11, where the CUD showed good performance. The difference between the highest and lowest CUC values was 24.8% between the four paddocks. With regard to CUD, the values obtained from the test were considerably lower than the CUC values in all four paddocks, representing an increase of 23.8% from the lowest to the highest CUD value.

Table 5 - Technical parameters relating to the current performance of the sprinkler irrigation system: Christiansen uniformity coefficient (CUC), distribution uniformity coefficient (CUD), statistical uniformity coefficient (CUE), irrigation efficiency (EI), water distribution loss (PDA), potential application efficiency (EAP) and evaporation losses (PV). Fortaleza - CE, 2016.

	CUC (%)	CUD (%)	CUE (%)	EI (%)	PDA (%)	EAP (%)	PV (%)
Picket 3	67,2	66,7	47,2	56,8	43,2	77,7	22,3
Picket 6	63,7	59,6	35,8	59,6	40,4	74,1	25,9
Picket 11	73,8	70,8	62,7	70,8	29,2	57,5	42,5

Picket 14	79,5	63,8	73,3	63,8	36,2	97,0	3,0

Source: Borges, R. C., 2016.

According to Mantovani (2001), the coefficients obtained when evaluating the performance of the irrigation system implemented in the paddocks evaluated showed reasonable and poor values, demonstrating the irrigation system's poor performance, according to the classification mentioned in Table 1. The water levels collected during the system's evaluation period varied greatly between the measured values (2 mm to 19.3 mm), as evidenced by the low coefficient values (Table 6).

Table 6 - Blades (mm) collected during the evaluation of the mesh sprinkler irrigation system. Fortaleza - CE, 2016.

	Picket 3			Picket 6		Picket 11		Picket 14				
4,8	5,4	12,1	7,4	4,2	2,6	2,6	2,1	2,0	4,8	3,6	5 2	4,3
4,0	5,0	6,2	7,0	5,8	4,0	5,6	4,2	3,8	5,2	3,4	6 4	6,6
4,2	3,8	4,8	5,4	6,2	5,1	3,7	3,6	4,8	2,4	3,6	6 3	6,0
5,7	4,8	9,0	19,3	3,1	4,8	3,6	3,2	4,4	2,1	5,2	6 1	4,8
5,2	5,8	6,4	7,2	10,4	8,0	7,9	3,4	4,1	3,8	5,0	6 0	5,5
4,0	4,0	7,7	6,2	4,6	6,7	3,6	3,0	3,8	4,2	3,4	5 5	5,2
4,4	2,2	11,7	2,8	5,0	5,6	2,0	2,4	2,4	3,0	4,0	5 0	4,8
						4,2	2,0	2,4	2,2	5,1	4 8	5,5
									2,2	5,6	6 6	6,3

Source: Borges, R. C., 2016.

According to Bernardo et al. (2009), sprinkler irrigation projects should have uniformity coefficients of around 80%, with the effect being a good spatial distribution of water and therefore greater uniformity of the pasture, which was not observed in the system evaluated. Similar results were also found by Bertossi et. al. (2013), who evaluated a mesh sprinkler irrigation system installed in an area of *Brachiaria brizantha* cv. Marandu pasture in Alegre-ES and obtained a large variation between the values of the collected blades, resulting in low CUC and CUD values.

Paulino et al. (2009) and Martins et al. (2011), when evaluating the performance of conventional sprinkler irrigation systems, also found low water application uniformity, attributing the low CUC and CUD values, as observed in this study.

In the case of CUE, according to the intervals defined by the ASAE (1996) shown in Table 2, paddocks 3 and 6 showed unacceptable values, 47.2% and 35.8% respectively. Pickets 11 and 14 showed a poor degree of acceptability (62.7% and 73.3%, respectively).

The wind speed during the evaluations of the sprinkler irrigation system was relatively moderate and may have had an effect on the uniformity coefficients. During the tests, wind speed values ranged from 0.87 m s^{-1} to 2.1 m s^{-1} . Bernardo et al. (2009) state that wind speed affects the distribution uniformity of sprinklers; the higher the wind speed, the lower the application uniformity.

When analyzing irrigation efficiency (IE), the following values were found: 56.8%, 59.6%, 70.8% and 63.8% in the respective paddocks 3, 6, 11 and 14. According to Bernardo et al. (2009) for the fixed conventional system, the values generally found are between 75% and 85% for efficiency, so the efficiencies found in the four paddocks are below those recommended in the literature, demonstrating poor performance of the irrigation system (Table 5), directly affecting the increase in water and energy consumption due to greater losses, thus increasing production costs (PAULINO et. al... 2009), 2009).

The values for evaporation losses (Pv) shown in Table 5 were 22.3%, 25.9% and 42.5% in paddocks 3, 6 and 11, which can be considered high. In paddock 14, the value found was 3%, considered low in relation to the other paddocks and corroborating the authors Beskow et al. (2008a) and Beskow et al. (2008b), who found values of water loss by evaporation and wind drag equal to 4.69% and values of less than 10%, respectively.

This evaporation loss seen in paddock 14 may be correlated with the weather conditions on the days the evaluations were carried out, as can be seen in table 7, depending on the period during which the work was carried out, corresponding to half an hour of system operation for each paddock.

Table 7 - Weather conditions observed on the days the tests were carried out, Fortaleza - CE, 2016.

Picket	**Date**	**Temperature Average (°C)**	**Wind speed ($m\ s^{-1}$) at 2m height**
3	24/11/2015	32,9	0,88
6	18/11/2015	32,0	0,87
11	02/11/2015	33,3	1,29
14	01/11/2015	31,2	2,13

Source: Borges, R. C., 2016.

Figura 7 - Graphical representation of the isopleths that represent the rainfall collected in the test for the area corresponding to picket number 3 obtained from the Surfer 11 software. Fortaleza - CE, 2016.

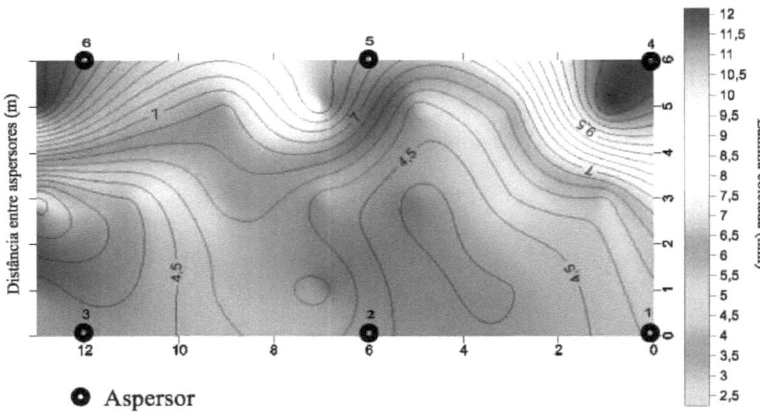

Source: Borges, R. C. (2016)

Figura 8 - Graphical representation of the isopleths that represent the rainfall collected in the test for the area corresponding to picket number 6 obtained from the Surfer 11 software. Fortaleza - CE, 2016.

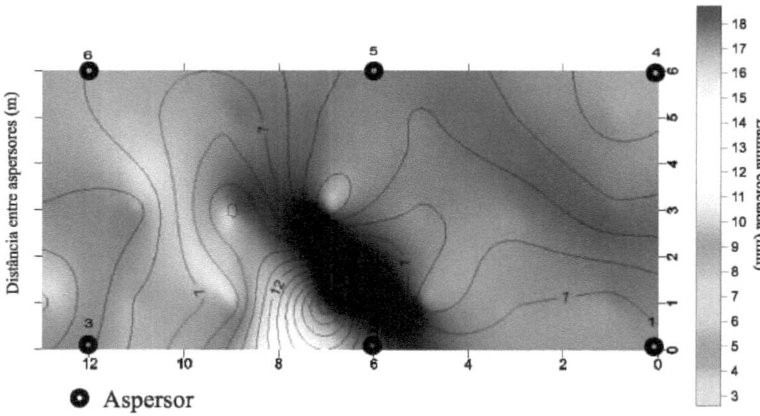

Source: Borges, R. C. (2016)

Figura 9 - Graphical representation of the isopleths that represent the rainfall collected in the test for the area corresponding to picket number 11 obtained from the

Surfer 11 software. Fortaleza - CE, 2016.

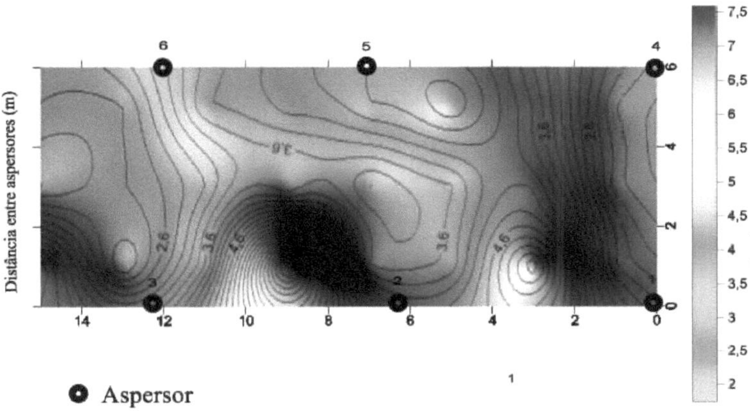

Source: Borges, R. C. (2016)

Figura 10 - Graphical representation of the isopleths representing the rainfall collected in the test for the area corresponding to picket number 14 obtained from the Surfer 11 software. Fortaleza - CE, 2016.

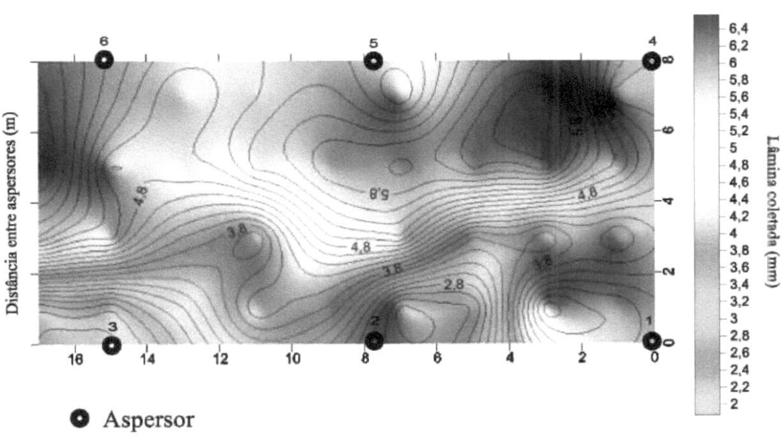

Source: Borges, R. C. (2016)

It should be noted that there was a considerable range of rainfall in all the paddocks, a fact which may explain the low values of the parameters identified in the evaluation of the system's performance, since during the test it was found that the sprinkler's water distribution was visibly influenced by the position of the sprinkler's adjustment deflector, having a direct influence on the amount of water applied and the reach of the water jet, corroborating the findings of Guirra (2011).

It can be seen in Figures 7, 8, 9 and 10 that there is a concentration of làmina at the points near the sprinklers, which can be explained by the fact that these sprinklers had the highest flow rates compared to the others. Another factor to consider in this poor distribution may be related to the sprinkler's operating pressure. The increase in pressure may have caused the water jet to spray more, which would have resulted in a reduction in the radius of reach. Bernardo et. al. (2009) report that a very high working pressure results in a high spray of the water jet, reducing its radius of reach and consequently causing a concentration of precipitation close to the sprinkler, while a very low pressure implies an inadequate spray of the water jet causing a very irregular distribution profile.

According to Siginer (1987), if irrigation is carried out unevenly, there will be areas that are over-irrigated, which in the pictures are the more intensely colored areas, and under-irrigated areas, shown by the lighter colors, where it is possible that the fresh matter productivity and, consequently, the dry matter productivity of the grass (brachiaria and tifton 85) is visibly different from the rest of the pasture.

Figures 11, 12, 13 and 14 show the graphical representations of the three-dimensional water application profile of the sprinklers installed in the areas corresponding to paddocks 3, 6, 11 and 14, obtained using Surfer 11 software.

Figura 11 - Graphical representation of the three-dimensional water application profile of the sprinklers installed in the area corresponding to picket number 3 obtained using Surfer 11 software. Fortaleza - CE, 2016.

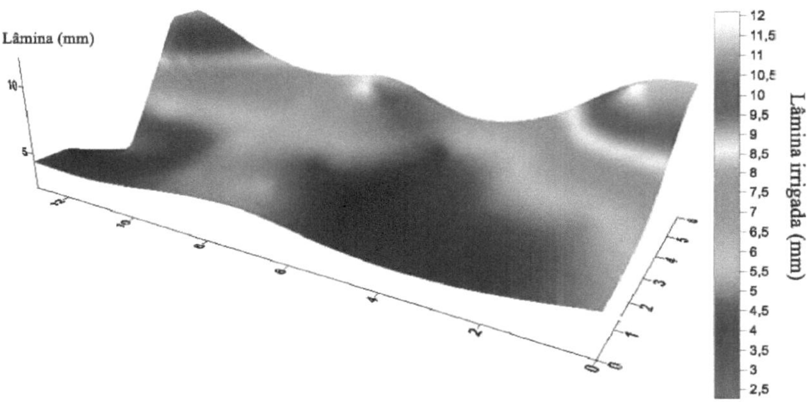

Source: Borges, R. C. (2016)

Figura 12 - Graphical representation of the three-dimensional water application profile of the sprinklers installed in the area corresponding to picket number 6 obtained using Surfer 11 software. Fortaleza - CE, 2016.

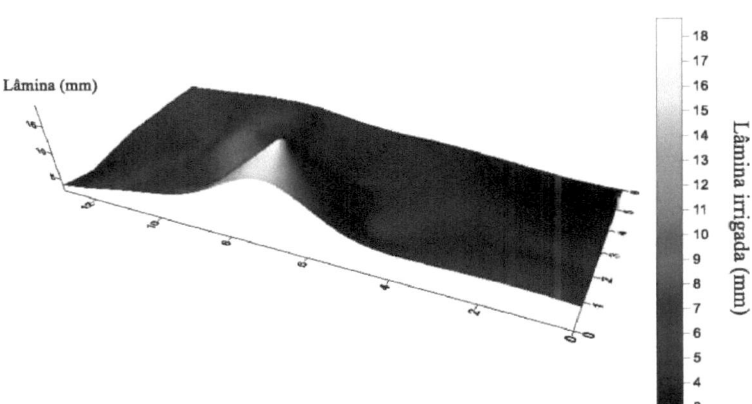

Source: Borges, R. C. (2016)

Figura 13 - Graphical representation of the three-dimensional water application profile of the sprinklers installed in the area corresponding to picket number 6 obtained using Surfer 11 software. Fortaleza - CE, 2016.

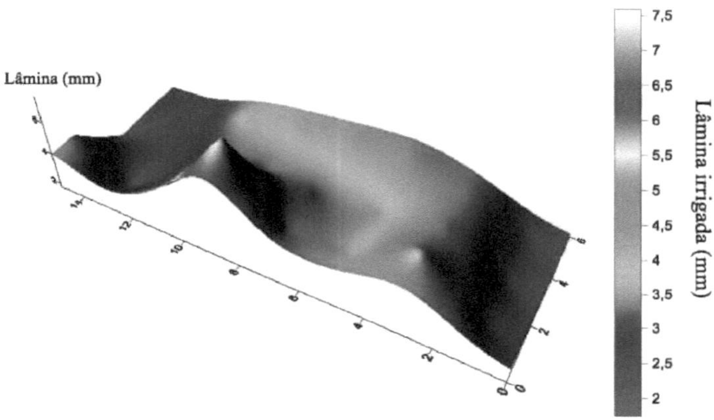

Source: Borges, R. C. (2016)

Figura 14 - Graphical representation of the three-dimensional water application profile of the sprinklers installed in the area corresponding to picket number 14 obtained using Surfer 11 software. Fortaleza - CE, 2016.

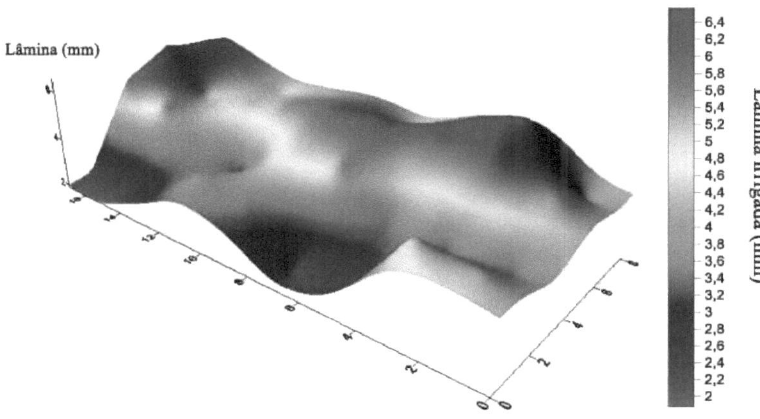

Source: Borges, R. C. (2016)

The low efficiency of sprinkler irrigation in field conditions has been frequently published in the literature, and these results are directly related to the significant unevenness of *water* distribution, as shown in the Figures

It is likely that the differences found between the peaks of water deficit and excess highlighted in the three-dimensional water application profiles may be related to the action of the winds during the time the test was carried out. Confirming these assumptions, Bernardo et. al. (2009) report that this low efficiency, in addition to being due to the uniformity of water distribution, is also related to water losses due to evaporation and wind drag, these losses being mainly affected by wind speed, relative humidity and air temperature.

With regard to this aspect, Christiansen (1942) describes that the effect of the wind can alter the model of water distribution by the sprinklers, resulting in a high concentration of water in the vicinity of the sprinkler, especially in the direction normal to the wind, thus causing a deficiency in the location opposite to the direction of the wind.

5. CONCLUSION

The irrigation project evaluated in the field showed poor uniformity of water application. The distribution of irrigation water in the areas of the paddocks under study is impaired, since its irrigation efficiency and all the uniformity coefficients evaluated showed values below those recommended in the literature. As a result of this poor distribution, the performance of the mesh sprinkler irrigation system evaluated is inadequate for this type of system.

6. REFERENCES

ANDRADE, C. L. Selection of the Irrigation System. **Circular Tècnica**, n.14. Minas Gerais: Ministério da Agricultura, Pecuâria e Abastecimento, 2001. Available at: < http : //docsagencia. cnptia.embrapa.br/milho/circul14.pdf>. Accessed on 04/01/2016.

ANDRADE, C. L. T; BRITO, R. A. L. Cultivation of Maize: Irrigation. **Technical Circular**, n.86. Sete Lagoas, MG. EMBRAPA Milho e Sorgo. 2006. Available at: <http://www.cnpms.embrapa.br/publicacoes/publica/2006/circular/Circ_86.pdf> Accessed on 17/02/2017.

ASAE - AMERICAN SOCIETY OF AGRICULTURAL ENGINEERS. **Field Evaluation of Microirrigation Systems**. St. Joseph. 1996.

ALENCAR, C. A. B.; CUNHA, F. F.; MARTINS, C. E.; CÓSER, A. C.; ROCHA, W. S. D.; ARAÙJO, R. A. S. Pasture irrigation: current situation and recommendations for use and management. **Revista Brasileira de Zootecnia**, v. 38, 2009 (Suplem. Especial).

BARRETO FILHO, A. A.; DANTAS NETO, J.; MATOS, J. A.; GOMES, E. M.; Performance of a micro-sprinkler irrigation system installed at field level **Revista Brasileira Engenharia Agricola e Ambiental,** v.4, n.3, p.309314, 2000.

BERNARDO, S.; SOARES, A. A.; MANTOVANI, E. C. **Manual de irrigaçâo**. 8. ed. 2nd Reprint. Viçosa: UFV, 2009. 596 p.

BERTOSSI, A. P.; MILEN, L. C.; HOTT, M. O.; RODRIGUES, R. R.; REIS, E. F. Evaluation of a mesh sprinkler irrigation system in pasture. **Revista Nucleus**, v.10, n.1,

2013.

BESKOW, S.; COLOMBO, A.; PEREIRA, G. M.; TAVEIRA, J. H. S.; RICARDO, C. M. Water losses by evaporation and drag in sprinkler irrigation under the climatic conditions of Lavras-MG, using medium-sized sprinklers. **Revista Irriga**, v.13, n. 1, p. 113-127, 2008a.

BESKOW, S.; COLOMBO, A.; RIBEIRO, M. S.; FERREIRA, L. S.; ROSSI, R. Simulation of water losses by evaporation and drag, in the NY-7 sprinkler (4.6mm x 4.0mm), in conventional sprinkler systems. **Revista Engenharia Agricola**, v. 28, n.3, p. 427-437, 2008b.

BILIBIO, C.; CARVALHO, J. A.; MARTINS, M.; REZENDE, F. C.; FREITAS, E. A.; GOMES, L. A. A. Vegetative and productive development of eggplant under different soil water tensions. **Revista Brasileira de Engenharia Agricola e Ambiental,** v. 14, n. 07, p. 730-735, 2010.

CARDOSO, G. C. Some practical factors in pasture irrigation. In: SIMPÒSIO DE PRODUÇÂO DE GADO DE CORTE, 2., 2001, Viçosa. **Proceedings**... Viçosa: Federal University of Viçosa, 2001. p. 243-260.

CARVALHO, I. R.; PELISSARI, G.; HANUS, A. D. ROSA, G. M. Water demand of crops of agronomic interest. **Enciclopédia Biosfera**, Centro Cientifico Conhecer -, v.9, n.17; p. 2013.

CHRISTIANSEN, J. E. **Irrigation by sprinkling**. Berkeley: California Agricultural Station. 1942. 124p. Bulletin, 670.

CRIDDLE, W.D.; DAVIS, S.; PAIR, C.H.; SHOCKLEY, D.G. **Methods for evaluating irrigation systems**. Washington DC: Soil Conservation Service - USDA. 24p. Agricultural Handbook, 82, 1956.

DRUMOND, L. C. D. **Application of swine wastewater by mesh sprinkling:** hydraulic performance and dry matter production of Tifton 85. 2003, 102f. Thesis (Doctorate in Agronomy) - Faculty of Agricultural and Veterinary Sciences, Paulista State University, Jaboticabal.

DRUMOND, L. C. D.; AGUIAR, A. P. A. **Irrigação de pastagem.** Uberaba: L. C. D. DRUMOND, 2005. 210p.

DRUMOND, L. C. D.; FERNANDES, A. L. T. **Mesh sprinkler irrigation for family coffee growing.** Uberaba: UNIUBE, 2004. 88p.

DRUMOND, L. C. D. Irrigation of pastures. In: SIMPÒSIO INTERNACIONAL DE PRODUÇÃO DE GADO DE CORTE, 2008, Viçosa. **Proceedings**... Viçosa: UFV, 2008. p. 307-320.

EMBRAPA - National Center for Soil Research. **Brazilian soil classification system.** 2ª Edition - EMBRAPA Solos, Rio de Janeiro 2006. 306p.

EUCLIDES, V. P. B.; VALLE, C. B.; MACEDO, M. C. M.; ALMEIDA, R. G.; MONTAGNER, D. B.; BARBOSA, R. A . "Brazilian scientific progress in pasture research during the first decade of XXI cen - tury", Revista Brasileira de Zootecnia, v. 39, p. 151 - 168, 2010.

FARIA, L. C.; COLOMBO, A.; OLIVEIRA, H. F. E.; PRAD O, G. Simulation of irrigation uniformity of conventional sprinkler systems operating under different wind conditions. **Revista Eng. Agric**., v.29, n.1, p.19-27, 2009.

GARCIA, F. C. H.; BEZERRA, F. M. L; FREITAS, C. A. S. Irrigation levels in the productive behavior of Formosa papaya in Chapada do Apodi, CE. **Revista Ciência Agronômica,** v. 38, n. 02, p. 136-141, 2007.

GOMES, H.P. **Engenharia de irrigaçâo**. Joao Pessoa: Ed. Universitâria/UFPB, 1994. 344p.

GUIRRA, A. P. P. M. **PERFORMANCE OF THE NAANDANJAIN 435 SPRINKLER AS A FUNCTION OF DEFLECTOR POSITION**. 2011. 67f Dissertation (Master's Degree in Agronomy - Soil Science). Faculty of Agricultural and Veterinary Sciences - Unesp, Jaboticabal Campus, Jaboticabal, 2011.

IBGE - BRAZILIAN INSTITUTE OF GEOGRAPHY AND STATISTICS.

Captured on Dec. 23, 2016. Online. Available at: http://seriesestatisticas.ibge.gov.br/lista tema.aspx?op=0&no=1

KING, B.A.; TARKALSON, D.D.; BJORNEBERG, D.L.; TABERNA, J.P. Planting System Effect on Yield Response of Russet Norkotah to Irrigation and Nitrogen under High Intensity Sprinkler Irrigation. **American Journal of Potato Research** , v. 88, n. 2, p. 121-134, 2011.

LUNA, N. R. S.; ANDRADE, E. M.; CRISÓSTOMO, L. A.; MEIRELES, A. C. M.; AQUINO, D. N. Nitrate and chloride dynamics in soil and groundwater quality in the Baixo Acaraù irrigation district, CE. **Revista Agro@mbiente On- line** , v.7, n.1, p. 3-

62, 2013.

MANTOVANI, E. C. **AVALIA:** Program for the Evaluation of Sprinkler and Localized Irrigation. Viçosa, MG: UFV, 2001.

MANTOVANI, E. C.; BERNARDO, S.; PALARETTI, L. F. **Irrigaçâo, principios e métodos**. 3.ed. current. expanded. Viçosa: UFV. 2009. 355p.

MARTINS, C. A. S.; REIS, E. F.; PASSOS, R. R.; GARCIA, G. O. Performance of conventional sprinkler irrigation systems on corn (Zea mays L.). **Idesia**, v. 29, v. 3. p. 65-74, 2011.

MENDONÇA, J.C.; SOUSA, E.F.; BERNARDO, S.; SUGAWARA, M. T.; PEÇANHA, A.L.; GOTTARDO, R.D. Determination of the cultural coefficient (Kc) of the bean plant (*Phaseolus vulgaris* L.), in Campos dos Goytacazes - RJ. **Revista Brasileira de Engenharia Agricola e Ambiental**, v.11, n.5, p.471-475, 2007.

OLIVEIRA, H. F. E.; COLOMBO, A.; FARIA, L. C. Modeling the effects of wind on the jet range dimensions of a hydraulic cannon. **Revista Brasileira de Engenharia Agricola e Ambiental**, v.13, p. 818-824, 2009.

PAULINO, M. A. O.; FIGUEIREDO, F. P.; FERNANDES, R. C.; MAIA, J. T. L. S.; GUILHERME, D. O.; BARBOSA, F. S. Evaluation of the uniformity and efficiency of water application in conventional sprinkler irrigation systems. **Revista Brasileira de Agricultura Irrigada**, v.3, n.2, p.48-54, 2009.

PAZ, V. P. S.; TEODORO, R. E. F.; MENDONÇA, F. C. Recursos hidricos, agricultura irrigada e meio ambiente. **Revista Brasileira de Engenharia Agricola e**

Ambiental, v. 4, n. 3, p. 465-473, 2000.

PRADO, G.; COLOMBO, A.; BIAGIONI, P. F. Angle of rotation and spacing between reels in self-propelled irrigation systems with the PLONA-RL400. **Revista Engenharia Agricola**, v.27, n.2, p. 346-353, 2007.

SIGINER, I. Spatial water distribution in sprinkler irrigation. **In: Hillel, D. ed. Advances in Irrigation. Academic Press.** 4: 119 - 165. 1987.

WILCOX, J. C.; SWAILES, G. E. Uniformity of water distribution by some under tree orchard sprinklers. **Scientific Agriculture**, v.27, n.11, p.565-583, 1947.

MIX
Papier aus verantwortungsvollen Quellen
Paper from responsible sources
FSC® C105338

Printed by Books on Demand GmbH, Norderstedt / Germany